If You Lived Beside the OCEAN

Gloria G. Adams

Slanted Ink

Stow, OH 44224

If You Lived Beside the OCEAN

Gloria G. Adams

ISBN: 978-1-7367688-6-0

Based on an earlier work by the author,
My Backyard is the Ocean

Images: Pixabay.com

Slanted Ink

Stow, OH 44224

For Jim,
who has always
loved the ocean.

If you lived beside the ocean, you might **SEE...**

a playful dolphin showing off his silly sea stunts,

Dolphins can only see forward, sideways, and down. They have to roll onto their backs in order to see above them.

A curvy-necked, pretty-in-pink flamingo,

Flamingos can sleep standing on one leg. A group of flamingos is called a flamboyance.

or even a pop-eyed crab
claw-churning the sand.

Crabs have 10 legs. They walk and run sideways so that one leg never gets in the way of another.

If you lived beside the ocean, you might **HEAR...**

the short, quick bark of a whiskered sea lion

Sea lions live on both land and water. They can walk on their flippers and sometimes even run.

screeching, squawking hungry seagulls,

Seagulls are one of the only kinds of birds that can drink salt water. They love to steal food from other seabirds.

or even the roar of waves inside a seashell.

Some shells are hollow inside. Outside noise bounces around inside the shell, making it sound like the roar of the ocean.

If you lived beside the ocean, you might **SMELL...**

a sweet-scented white jasmine bouquet,

Jasmine flowers bloom at night. They are a favorite ingredient in such things as perfume, tea, and rice.

crisp, salty air as sea spray
smacks the rocks,

Salt in the ocean comes from the mineral salts that wash off rocks and also from openings in the sea floor.

or even a fishy whiff of the catch of the day.

Scientists have discovered that counting the scales on a fish can tell how old it is. Some fish can live to be 100 years old!

If you lived beside the ocean, you might TASTE...

crispy, bright green seaweed salad

Seaweed provides many vitamins and minerals, not only for people, but for marine animals, such as fish, manatees, and sea turtles.

the tangy taste of beach plum jam,

The beach plum plant is native to North America. It grows best in sandy places, like coastal sand dunes.

or even a bite of freshly grilled shrimp.

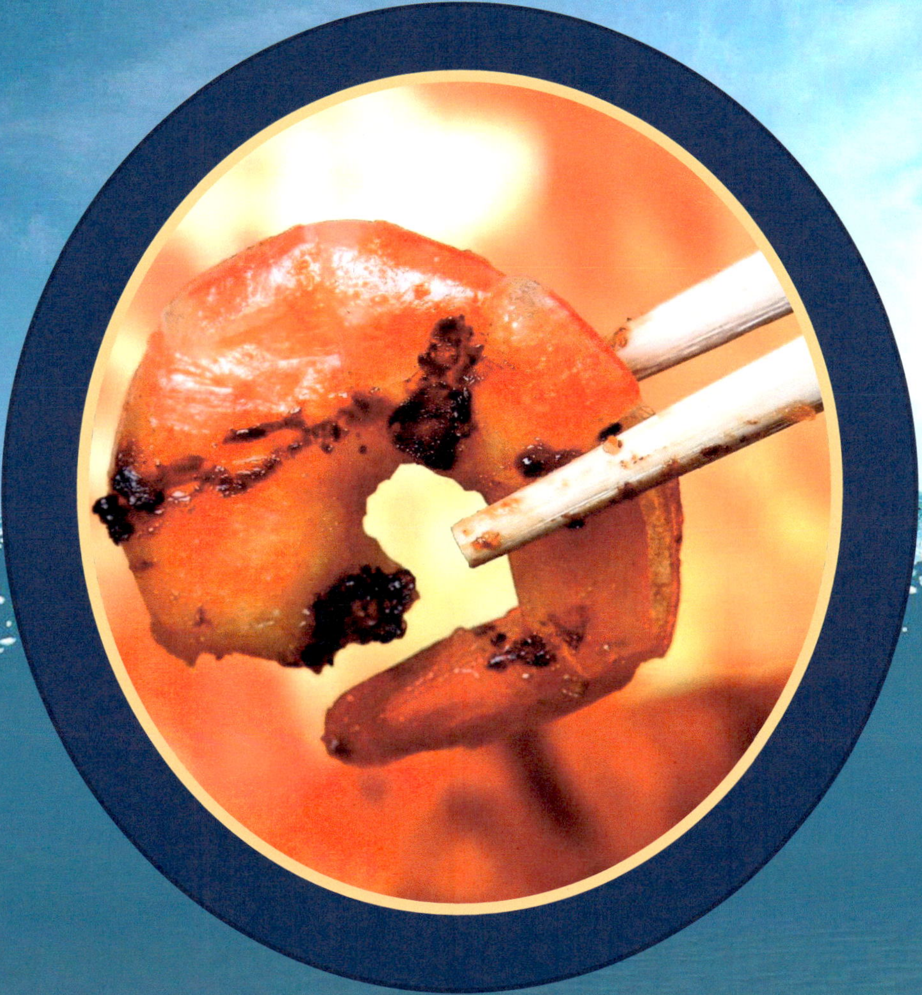

Most shrimp are small, but some species can grow up to 12 inches, as long as a ruler. In the ocean, shrimp can live for as long as six years.

If you lived beside the ocean, you might TOUCH...

smooth stone stacks
tumbled by the sea,

Rocks become smooth in a process called "weathering." The action of the waves rounds out the jagged edges .

cool, wet sand
between your toes,

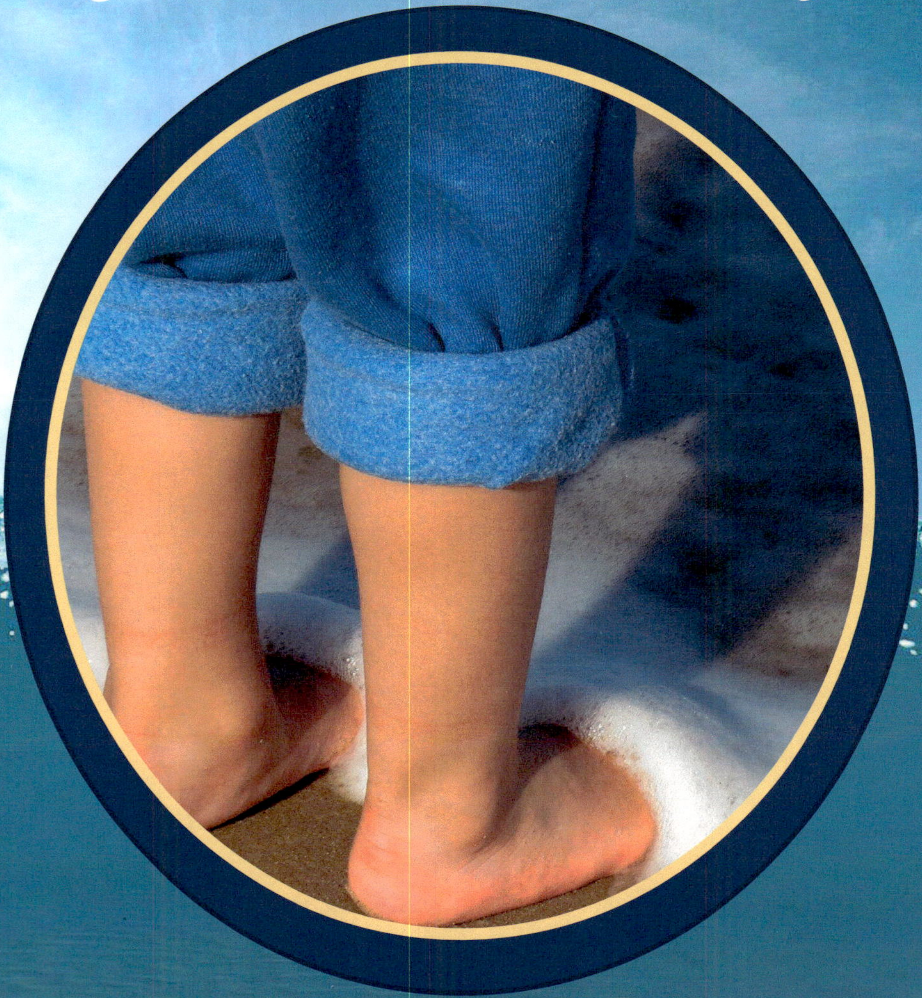

Seafoam is sometimes called "spume." One reason it forms is from the constant movement, called agitation, of ocean waves.

or even the rugged ridges of snow-white seashells.

In the ocean, two sea scallop shells are connected. They zoom through the water by clapping their two shells together.

And when the day ends,
you might even get to watch
the sun sink into the sea,

if you lived beside the ocean.

FUN FACT!

If you could gather all the salt from the oceans, it would cover the entire surface of our planet.

There are 5 oceans on our planet. They cover about 71% of the Earth's surface.

Where on the Earth do YOU live? What do you see, hear, smell, taste and touch in YOUR world?

Did You Know?

- Oceans are part of the Earth's **Aquatic Marine Biome**. A biome is an area of the Earth that has the same climate, plants, animals, insects, and geology.

- The **Marine Biome** includes all the oceans on the Earth.

- There is salt in ocean water, but not in lakes, ponds, rivers or streams.

- Mountains aren't only found on land. The biggest mountain range in the world runs down the middle of the Atlantic Ocean.

Lionfish

- The plants that grow in the ocean, called algae, supply about half of the oxygen we breathe.

- The biggest animal in the world lives in the ocean. It is the Antarctic Blue Whale and weighs about as much as 30 elephants!

- Water at the bottom of the ocean is hot, not cold.

- All the birds, plants and animals in this book are part of the **Aquatic Marine Biome.**

About the Author

I am a former children's librarian and storyteller. I believe that reading to children and helping them become strong readers is one of the most important things we can do. Visit my website at www.gloriagadams.com.

~ Gloria G. Adams

Other books by the Author

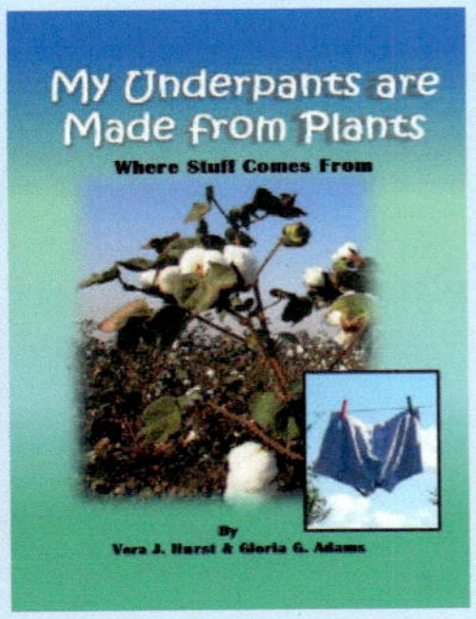

Made in the USA
Monee, IL
05 November 2026

67985535R00021